GERMAN ROCKET LAUNCHERS

Joachim Engelmann

The cover picture in color, based on a drawing by Hans Liska, shows a salvo fired by a 21cm launcher battery in the winter of 1943-44.
The photo on the inside cover gives an impression of a night firing of 21cm Smokelayer 42 rockets.

1469 Morstein Road, West Chester, Pennsylvania 19380

Photo sources:
Federal Archives, Koblenz
Engelmann Private Archives
Scheibert Archives
Podzun Publishers Archives

Translated from the German by Dr. Edward Force,
Central Connecticut State University.

Library of Congress Catalog Number: 90-60479.

Printed in the United States of America.
ISBN: 0-88740-240-2

This book originally published under the title,
Deutsche Raketen-Werfer,
by Podzun-Pallas Verlag, 6360 Friedberg 3 (Dorheim),
© 1977. ISBN: 3-7909-0057-5.

We are interested in hearing from authors with
book ideas on related topics.

A 15cm armored smokelayer (tenfold) on the "Mule." It has just been loaded and prepared for firing.

ROCKET LAUNCHERS

The wartime use of chemical war materials, incendiaries and rockets is ancient: The ancient Chinese used them, as did the Spartans in the Peloponnesian War, 431-404 B.C. In the 19th Century, England, Denmark, Switzerland, France (in Algeria) Greece (as of 1840), the USA (as of 1846), Prussia (until 1872), Austria (until 1867) and above all Russia, against the Poles, Turks, Chinese and in the Crimean War, used rockets as effective weapons. Their further development in Germany, on the other hand, was blocked by the improvement of artillery, until the course of World War I as of April 22, 1915, in the course of trench warfare and the use of chemical weapons, indicated a need for large-caliber mobile mine and gas launchers with long ranges, and the Treaty of Versailles, banning as it did heavy artillery, tanks and warplanes, put more and more emphasis on the strengthening of non-banned weapons like smoke screens and rockets. Opportunities thwarted since the middle of the past century were taken up, and the military cooperation that existed for a time with the "Red Army" under the Treaty of Rapallo led to development in many areas.

Ever since 1910 the rocket scientists Hermann Oberth and Goddard had carried on basic research which was followed by the first experiments with powder rockets in 1925, Fritz von Opel's and M. Valier's rocket cars in 1928, the test flight of a rocket plane in 1929 and Winkler's liquid-fuel rocket in 1931. Thus the first measures taken by the German Army and the Army Weapons Office under General of the Artillery Prof. Dr. Becker were not only timely and purposeful, but also showed great forward vision, as they were, in 1929-30, the first signs pointing to a new age, which was then characterized by Walter Dornberger and Wernher von Braun. Here Germany had an advantage of six to ten years. England began only in July of 1936 and was able to set up rocket batteries at the end of 1941; the USA followed as of 1939 and used rockets on ships and aircraft as of 1942. The Soviet Union entered the picture at the end of 1941 with their 82, 132 and 400mm "Katyushka" launchers, using steel beams on trucks and armored vehicles ("Stalin Organs"). Until 1945, though, Germany held a very commanding lead in the stormy further development of this weapon.

At first, proceeding from the smoke-screen and decontamination units, field rocket units were set up — within the framework of the developing smoke troops - in which the technical breakthrough was made under the demands of war, so that the emphasis was then: 1934, still smoke and grenade launchers; 1939, breakthrough to large rockets; developed 1942-45 to long-range rockets in the operative-strategic realm. Along with this, infantry-antitank rockets as of 1941 and anti-aircraft rockets as of 1942 appeared as a kind of by-product of this development. Artillery and chemical troops became launcher troops, represented to this day by artillery rocket and strategic rocket troops, aside from anti-aircraft and infantry rocket units. This technical breakthrough was based on the use of jet powerplants with electric ignition, which — free from static and dynamic ignition — with their freedom from recoil and their smooth barrels, allowed not only larger calibers but also new initial velocity ranges through their unlimited boost intensification — practically independently of the mantelet —allowed hitherto unknown ranges which turned the shell into an aircraft. With that the move through the atmosphere beyond the earth's gravity became possible, combined with electronic steering and automatic target seeking. Since 1945 the USA and the Soviet Union made use of the German designers and technicians.

After the development of spinning jets by the Army Weapons Office as of 1931, which used exit openings drilled at angles and rim jet ignition to transform the movement of the launched shell into a rotating motion around its longitudinal axis and thus, without wings — like the Russians — and without rifling — like the artillery — stabilized the direction of flight, and after the successful conclusion of experiments with tracer equipment by Hauptmann Dr. Ing. Walter Dornberger and his colleague Wernher von Braun at the end of 1935, succeeded by 1937, with the 15cm NbWf 41, in developing the standard multiple launcher, which not only replaced all previous types but was manufactured and used until the end of the war. The "Do-Werfer" offered the following advantages: Its freedom from recoil allowed a low launcher weight (540 to 1200 kg), representing only about 1/3 to 1/10 of that of corresponding artillery guns, a low firing elevation, and a favorable potential for off-road use, even capable of being moved by manpower when necessary.

The weapon was robust, long-lasting and simple, simple to operate as well. The smooth barrels showed practically no wear; the launching apparatus was standardized to a great degree; most launchers took only one type of load. Thus the launcher cost only 1/3 to 1/20 of a comparable artillery gun (1500 to 5000 Reichsmark).

Thus the launcher is a "poor man's" low-cost weapon. And its production cost only 1/2 to 1/3 as much as the corresponding artillery weapons! The heavy launching apparatus and launching frames even fired their projectiles right out of the packing case.

Not only the (then) large calibers of the projectiles and shells, of 15 to 32cm, but particularly the doubled to equal shot weight (of the large calibers), the doubled to ninefold amounts of explosives in comparison to the shot weight, and the great shock wave through the thin walls of the shell with low shrapnel effect, added up to an extremely heightened effect on the target. The great shot weights were equal to those of the 21cm mortar. Another factor was the area covered. A launcher battery covered more than the effective area of an artillery unit with 650 by 100mm, a launcher unit more than that of an artillery regiment with approximately 2000 by 100 meters. An effect of this kind was unknown until then. Electric ignition of the launchers in firing increased the rate of fire; the mixture of explosive, incendiary and smokelaying ammunition intensified the effect.

As army troops, the launchers were fully motorized; a partial organization as divisional artillery regiments from March 1, 1942 to 1943 — as, for example, into A.R. 336, 340, 345, 371, 376, 377, 383, 384, 386 (mot.), 387, 389 and the 7th and 8th units of the SS artillery batteries, whether launcher batteries or launcher units — remained an exception. Launchers and launcher units had to be particularly mobile, because after their first salvoes their positions were rather easy for the enemy to locate, so that they had to be moved again at once in order to avoid losses. Therefore they were mounted on or towed by rather large cross-country vehicles such as 1-ton and 3-ton halftrack towing tractors.

Despite the few disadvantages of the weapon — half the range of the artillery, no utility as individual weapons, no effect on pinpoint targets, little destructive power against emplacements and structures, no barrage fire just in front of the front line — the launcher troops were regarded as focal-point weapons, used to open the way for some large-scale attacks and, in defensive action, to smash enemy emplacements and smash attempted breakthroughs. They combined the surprise factor of their sudden attacking fire with superior and decisive intensity of impact as a mass weapon. This required constant planning ahead by their leadership and particular consideration for deployment, unlike the conventional artillery, even though the two had to cooperate for total effect. The use of one unit was the rule, that of one battery the exception, even though its was sufficient; larger concentrations were used only in large-scale combat. This is why the Army Weapons Office had decided in favor of multiple launchers with few (4 to 10) barrels, but many of them, as opposed to the Russians with their multiple launcher units.

In ten years the development proceeded more than stormily from the first experimental units to powerful launcher troops that attained their full importance only as of 1941 and experienced their greatest total use in mid-December of 1944. They had increased fiftyfold in six years, between 1939 and 1945; in 1939 they represented the newest and smallest but one of the most modern service arms, with 100 officers, 332 non-commissioned officers and 1612 enlisted men, a total of 2044 men in all, with 96 launchers and 484 motor vehicles. In 1945, on the other hand, they numbered 5257 officers, 18,150 non-coms and 88,914 enlisted men, a total of 112,321 soldiers, with 4816 launchers and 27,066 vehicles! The production of launchers and ammunition from 1941 to 1945 was as follows:

Launchers

	1941	1942	1943	1944	1945 (Jan.)	
15 cm Nb.Wf. 41	650	970	1.188	2.336	139	= 5.283
21 cm Nb.Wf. 42	648	970	100	835	73	= 2.626
30 cm Nb.Wf. 42	-	-	380	544	30	= 954
Totals	1.298	1.940	1.668	3.715	242	= 8.863

Ammunition (by thousands)

15 cm NB.Wf. 41	418	1.208	1.096	1.985	120
21 cm Nb.Wf. 42	-	9	120	258	12
28/32 cm NbWf 41	125	169	143	140	4
30 cm Nb.Wf.42	-	24	106	145	6
Totals:	543	1.410	1.465	2.528	142

When the war ended in 1945 the launchers, about half of their numbers lost, had handled important artillery assignments. Other tasks were performed by the long-range rockets, the V1 (for ten months) and V2 (for seven months), the infantry's antitank rockets like the Panzerfaust, Ofenrohr, Panzerschreck and Puppchen, and the anti-aircraft rockets such as the Schmetterling, Enzian, Rheintochter and Wasserfall. All of them contributed to an epoch-making era.

In July of 1944 Hitler demanded a new high level of production, with an increase to 3.6 million rounds of launcher ammunition; on August 20, 1944 he urged an immediate "push" for 1500 new launchers as of October 20 of the same year, and this figure was reached in a tremendous burst of five to seven months' production. Thus on December 16, 1944, at the beginning of the Ardennes offensive, eight launcher brigades with 957 launchers, including 369 heavy ones, could see action in three attacking armies, the brigades being, per army corps:

6th Panzer Army: Launcher Brigades 4, 9 and 17, with 214 Nb.Wf. 15cm, 108 Nb.Wf. 21cm, 18 Nb.Wf. 30cm, in all 214 light and 126 heavy launchers = 340 launchers.
5th Panzer Army: Launcher Brigades 7, 15 and 16, with 232 Nb.Wf. 15cm, 54 Nb.Wf. 21cm, 54 Nb.Wf. 30cm, in all 232 light and 108 heavy launchers = 340 launchers.
7th Army: Launcher Brigades 8 and 18 with 140 Nb.Wf. 15cm, 54 Nb.Wf. 21cm, 54 Nb.Wf. 30cm, in all 140 light and 108 heavy launchers = 248 launchers.

This added up to 3528 15cm, 1215 21cm and 756 30cm barrels in the army launcher troops alone! Aside from the final failure brought about by the Russian large-scale attack in the east, the weather, and enemy air superiority, this last large-scale German attack was the high point of the German rocket launcher. For the first time, the number of launcher barrels exceeded that of the artillery guns with 1003 light and 659 heavy launchers, making more than twice as many as guns.

10cm NbWf 40

The first 10cm Nebel-Werfer 35 was a very simple lightweight design that could be moved by its team in three loads or on launcher cars. In exceptional cases it could be carried on a special vehicle along with its crew. It was in practice nothing more than a minethrower, cheap, functional and durable, but with modest range and effect.

An improved successor design, the 10cm Nb.Wf. 40, more than doubled the range with its barrel lengthened by some 50cm, but made the launcher eight times heavier and nine times more expensive to produce, had only 3/4 of the former rate of fire and suffered three times as much wear. The massive carriage had no relation to the performance of the weapon. Even though its move to a motor vehicle brought a touch of modernization, that could not be regarded as satisfactory progress!

A battery during launcher drill in Germany. The soldiers are members of the ***Hermann Göring*** Regiment and therefore are wearing Luftwaffe uniforms.

The launcher is ready to fire; the crew awaits the leader's command to fire at any moment.

Above: "Fire!" - in fractions of a second the gunner sends the shell on its way out of the long launcher barrel, which measures almost two meters and can be raised to an elevation of 84 degrees.

10cm Wgr. 35 Nb.

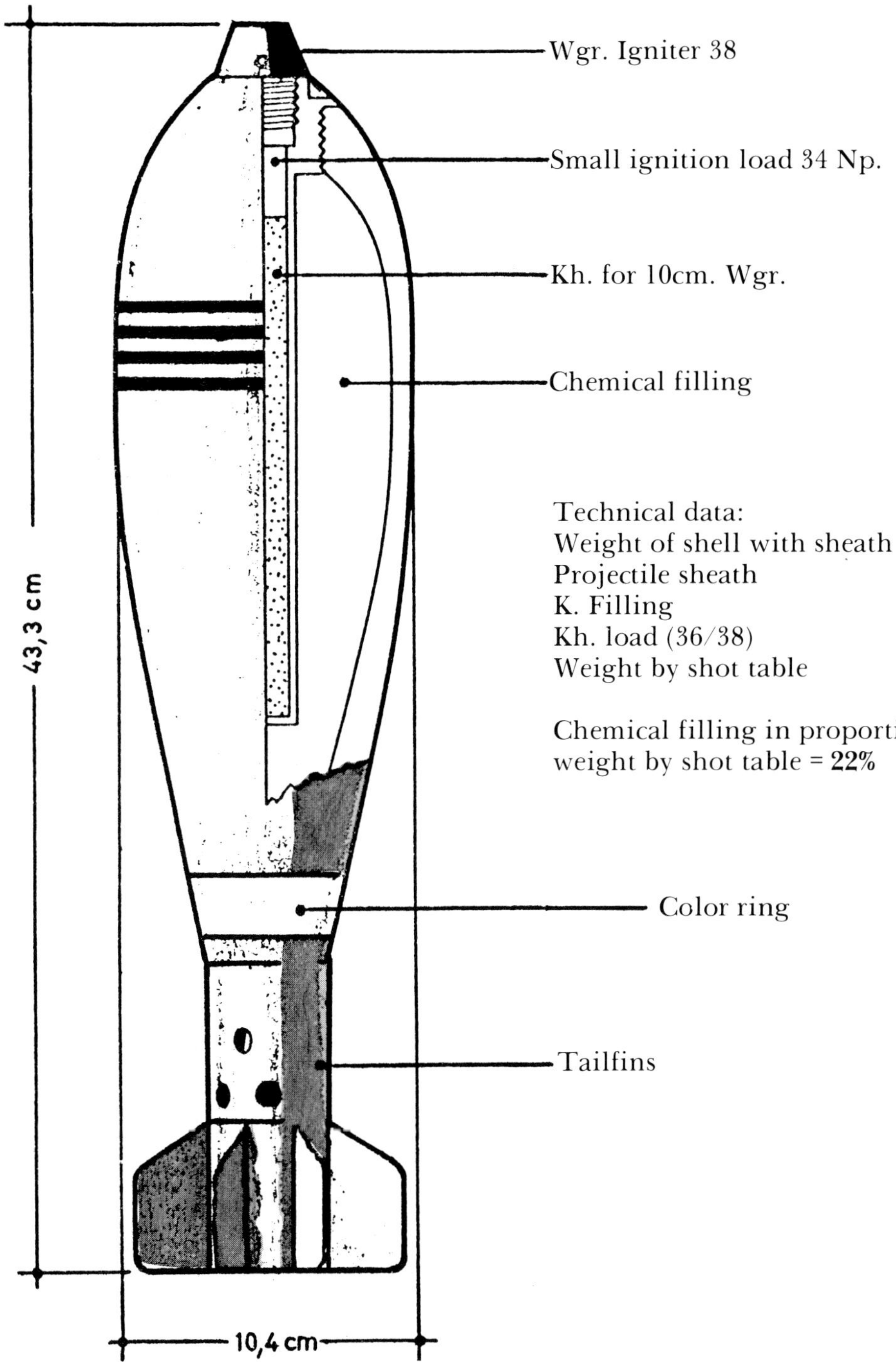

Wgr. Igniter 38

Small ignition load 34 Np.

Kh. for 10cm. Wgr.

Chemical filling

Color ring

Tailfins

43,3 cm

10,4 cm

Technical data:	in kg
Weight of shell with sheath	8.2
Projectile sheath	1.4
K. Filling	1.5
Kh. load (36/38)	0.05
Weight by shot table	6.8

Chemical filling in proportion to weight by shot table = 22%

The 10cm Nb.Wf. 35 launcher shot six different types of mortar shells: 10cm Wgr. 35 Nb.St., Nb.Te. and St., Spr., Nb.(Ub.), Br. and 10cm Wgr 37. Like the 10cm Nb.Wf. 40 launcher, it could take three charges which, when used in these two types of launcher, reached an initial velocity between 193 and 310 meters per second, stabilized by six stub tailfins like traditional shells. More than one-fifth of these shells could be filled with chemicals.

But neither the caliber nor the shot weight was actually sufficient for effective use. The decisive improvement needed rather to be sought by changing from a single to a multiple launcher and continuing to develop the shell propellant from a simple powder charge to a non-recoil rocket type, in the process of which the thus intensified problem of shot stabilization had to be solved. The future launcher could then have a lighter carriage and no longer needed to cost more than three or four times that of the original type 35 in terms of materials, expenses or production time.

The transition to rocket drive simultaneously allowed a significant increase in caliber, a large increase in shot weight and a significant increase in shot range. The German rocket shells were, to be sure, not stabilized by fins, but rather by diagonal-jet spinning, in order to diminish going off-course.

11cm Tracer Device

After black-powder rockets made by the firm of W. Sander in Wesermünde were used successfully to propel racing cars, gliders and sleds,the Army Weapons Office decided at the end of the Twenties to develop jet-propelled weapons in Kummersdorf, near Zossen, under the direction of Hauptmann Dipl.Ing. Ritter von Horstig and Hauptmann Dr. Ing. Walter Dornberger as expert advisor and specialist in rockets (Wa.Prf.11). Three and a half years of experiments allowed time to set up specifications and establish shot tables, until at the end of 1934 a troop test of 11cm black-powder rockets by the 2./Art.Abt. "Königsbrück" began, concluding successfully for the first time at Jüterbog in 1935.

This forerunner of the Do-Werfer consisted of four rails 3.2 meters long, held together by belts. At the rear was a firing pan to hold the 11cm powder rocket. The firing track was anchored at the rear by a stake in the ground and screwed onto two legs at the front, which provided an aiming field of elevation and traverse similar to that of the 10cm Nb.Wf.35; a range of 4500 meters was attained, and batteries of two columns, each with nine quadruple launchers, were planned.

Ignition was provided by Rim Jet Igniter 34 and an electric igniting machine. The tests had indicated a probability of fin-stabilized rockets going off course which was too great for military use. For that reason in 1931 the rocket experts of the Army Weapons Office designed spinning jets to keep the rockets on course. By changing the nucleus of powder, speeding up the burning and enlarging the caliber as well as changing to smokeless powder, the range could be increased, the payload enlarged and the accuracy improved.

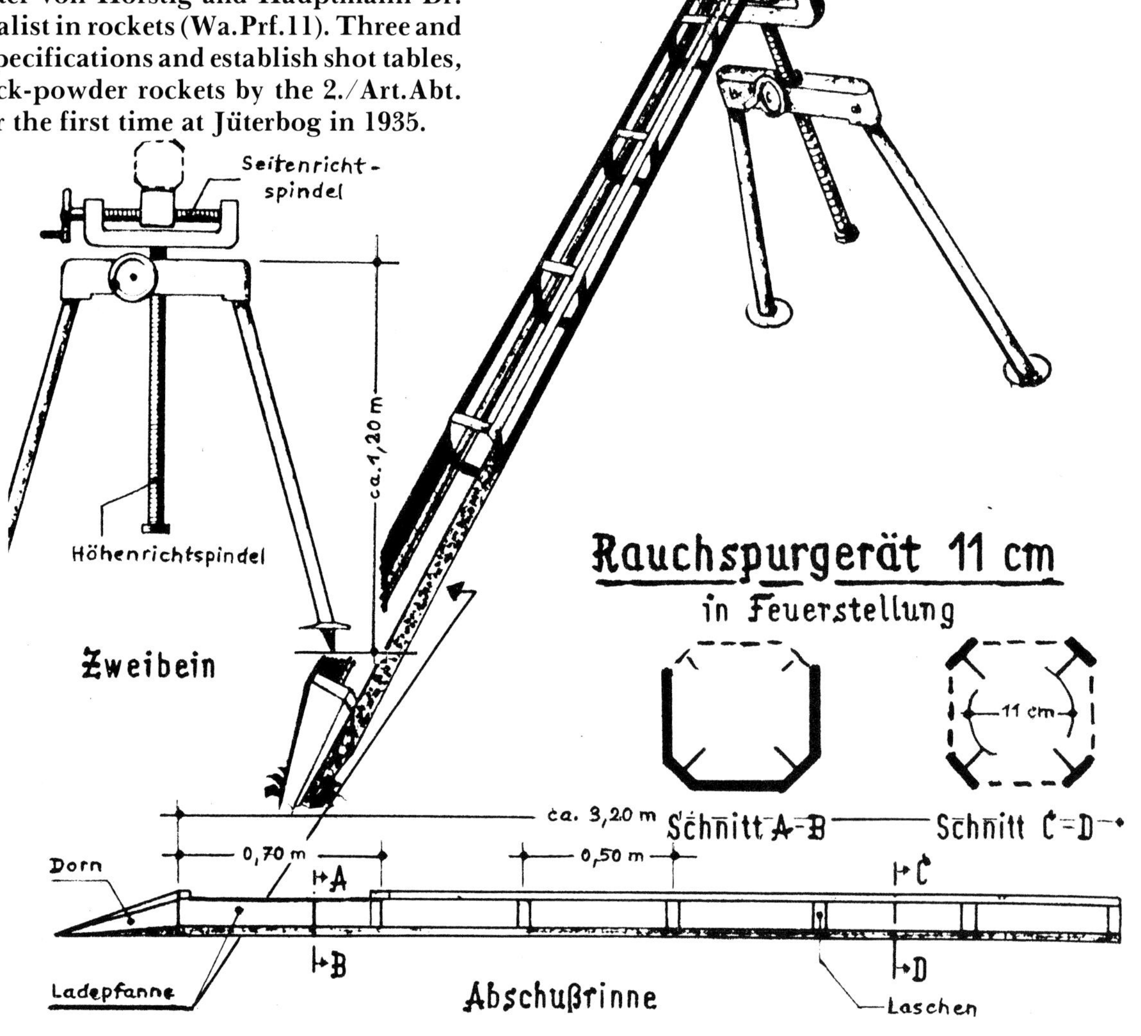

(Richtaufsatz:) Aiming gear
(Seitenrichtspindel:) Traverse aiming spindle
(Höhenrichtspindel:) Elevation aiming spindle
(Zweibein:) Two-legged mount
(Dorn:) Stake
(Ladepfanne:) Loading pan
(Abschussrinne:) Firing track
(Laschen:) Belts

Tracer Device 11cm

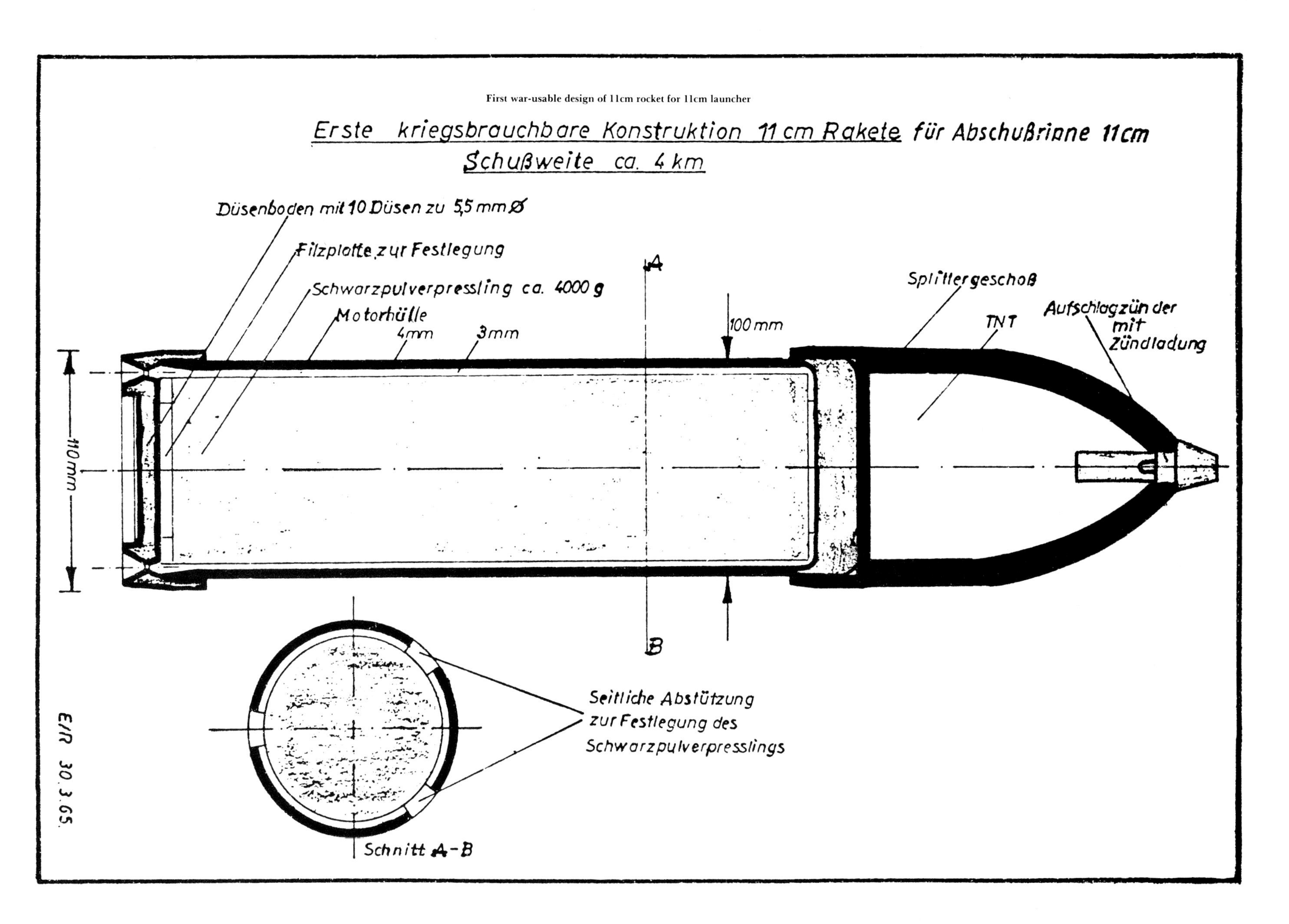

With this rocket the technical breakthrough into a new age of weapons was attained. The very escalation of the technical factors opened unsuspected potentialities.

Generalmajor Dr. Ing.h.c. Walter R. Dornberger, born in Giessen on September 6, 1895, was a brilliant designer who not only created the German launchers technically but also created the rocket weapon — an unusual career in the German Army. He joined Foot Artillery Regiment 3 Brandenburg in August of 1914, reached the rank of Leutnant by July of 1915 and was captured by the French in October of 1918. In the 6./A.R.6 of the Reichswehr he worked with Lt. Froben to improve sound and light measuring technology and troop surveying and did pioneering work in motor racing.

After passing the "Technical Defense examination" with a high score, he was sent in 1928 by his unit commander, Major von Brauchitsch, to study at the Technical Institute in Berlin-Charlottenburg, where he received his diploma in engineering with honors in 1930. Beginning in 1938, Wernher von Braun, with his outstanding organizational and technical capabilities, became his closest collaborator and remained so for seven years. To close technical gaps in the defensive power of the Reichswehr, Dornberger, with the advice of Dr. Oberth, tested ideas and designs of civilian inventors at the "Berlin Rocket Airfield." Practical troop tests as Battery Chief of the 2./Art.Abt. Königsbrück, beginning in 1934, brought positive results. In 1935 a secret study of rocket technology won him his degree of Dr.Ing.h.c.

Just one year later, Gen.Maj. Prof. Dr. Becker summoned Hauptmann Dornberger to the Army Weapons Office as Department Chief for Rocket Development (Wa.Prüf.11). Rocket development was then taken forward regularly at the Kummersdorf Firing Range, until his engineering group created the first practical rocket salvo launcher, the 15cm Do-Werfer 41. In 1939, at the rank of Oberstleutnant, he was also entrusted with the leadership of the Army Test Center at Peenemünde, which later developed into the "Rocket Test Center", the independent developmental facility of the "V-Weapon Program of the Army." In October 1942 he experienced the first launching of the V2, and on July 7, 1943, at the rank of Generalmajor, became the "Deputy for Special Experimentation." After many difficulties he was finally able on September 15, 1943 to attain for Peenemünde the status of "Special Urgency", almost too late. On June 13, 1944 the massive use of the V1 began, and as of September 6, 1944 that of the V2 (A 4). At the end of the war, the V3 (A 9/10) intercontinental rocket was finished in terms of design, but was never used. These achievements earned him the Knight's Cross of the War Service Cross.

After the war ended, Dornberger and von Braun were taken to the USA, where, unlike Germany and Russia, there was no military development of large rockets. The present-day American rocket weapon is based on the principles of Dornberger's engineering group. After several years of US Army service, Dornberger became the Vice-President of the Research Department of the Bell Aero Systems Company in Buffalo, New York and gained official recognition late in life when he received the Eugene Sänger Medal.

His closest collaborators were Gen. Maj. Dipl.Ing. Leo Zanssen, Oberst Dipl.Ing. Sommerkorn, as well as Dr. Schemel, Reg. Baurat Dr. Pöppl, Reg. Baurat Dipl.Ing. Pitzken, Reg. Ob.Baurat Dr. H. Schwab and Gen. Maj. Dipl.Ing. Josef Rossmann as Chief of the Rocket Group in the Armament Office. Inspectors of the Smoke Screen Troops were, from 1936 to 1939, Generalleutnant Theisen, Generalmajor von Tempelhoff from 1939 to 1941, Generalleutnant Leister 1942-44, and Generalmajor von Blücher 1944-45, all of whom naturally also played a role in weapon development.

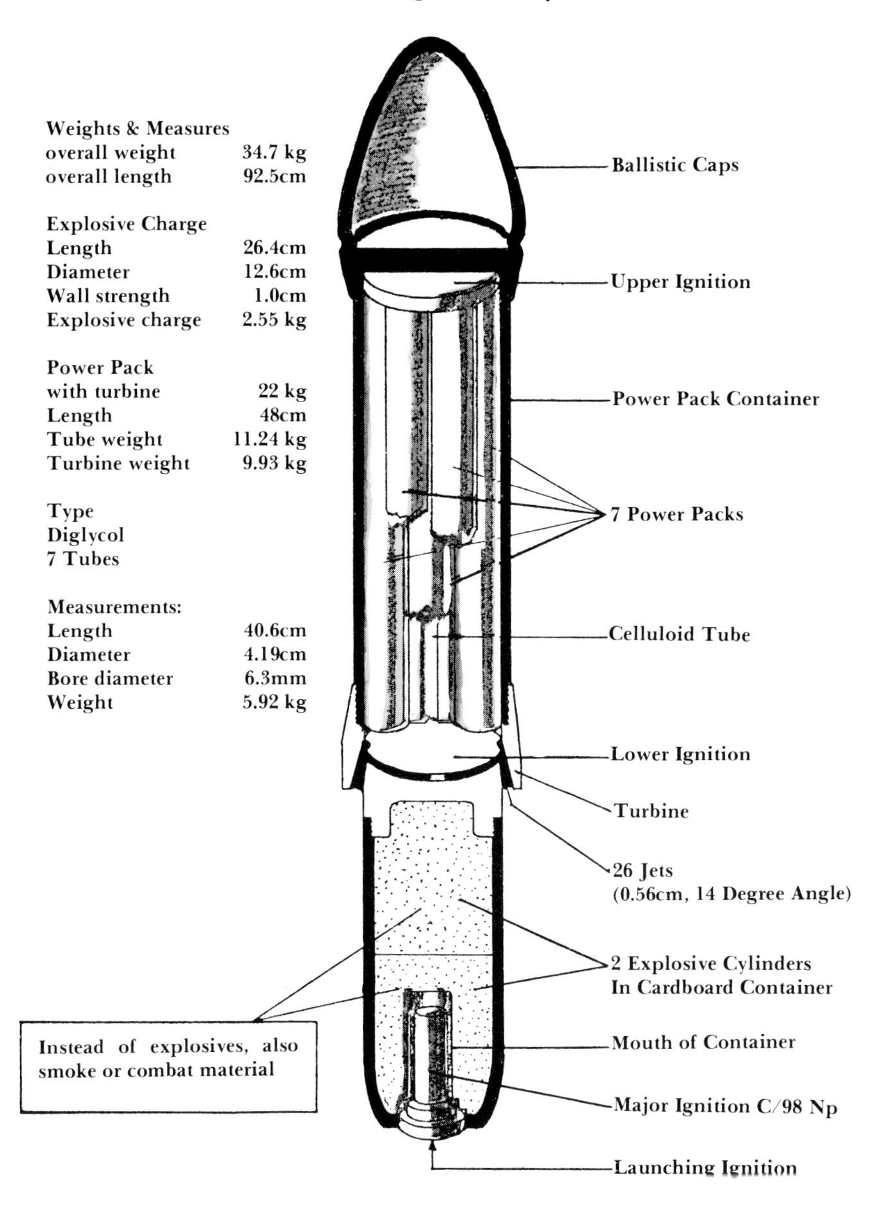

15cm NbWf 41

The 15cm Nb.Wf. 41 set up and ready to fire but not loaded, seen in front 3/4 view on a Michten spreading carriage and mount (wheels off the ground), easy to see here with its bottom plate.

Left: The sixth barrel of a 15cm Nb.Wf.41 is loaded. The ignition distributor of the electric rim jet igniter (ERZ 39) with its connection to each individual barrel is clearly visible. The shell is pushed into the barrel up to the turbine assembly, so that the 2.55 kg power pack with the ignition charge C/98 Np. projects out the rear.

This launcher was produced at the Frama Works in Hainichen and the Saxon Textile Machine Factory in Chemnitz.

To avoid the term "rocket" the designations "smoke-screen layer" or "Do-Werfer" (after Dornberger) were used; the name of Dipl.Ing. Rudolf Nebel has — as has often been claimed — no connection.

Right: A cannoneer ignites the launcher from a distance of 30 meters, using the electric sixfold igniting machine via the sevenfold ignition cable. Three projectiles with diglycol propellants are in flight. The unit carried 20 salvos of explosives and 5 salvos of smoke.

The launcher fires explosive and smoke shells in normal, tropic and arctic versions for black powder, diglycol and triglycol. It is not a single-function weapon.

Right: A 15cm Nb.Wf. battery firing; the burning of the black-powder propellant charge produces a mighty cloud of smoke.

21cm Nb.Wf. 42

The power of the mighty Soviet launcher weapon "Katyushka" ("Stalin Organ"), increased as of March 1942 and consisting of 38 truck-mounted barrels made of best-quality steel charged with nitroglycerine and powered by a central jet instead of a turbine, with a range of 5150 meters but lesser effect, as the explosive charge was in the warhead, pressed the German leadership to make increased use of launchers and develop an improved launcher design. Thus the 21cm Nb.Wf.42 was developed as a standard medium launcher.

The 550 kilogram launcher is put in position by its crew. The bottom plate can be seen under the mouths of the barrels.

The medium standard launcher — made at the Donauwörth Machine Factory — strengthened the already existing launcher regiments as of 1942and made the formation of focal points within these groups possible. The unit, numbering 575 men, carried as its primary armament 900 explosive shells, 180 of them for each battery, 360 for the lead column. This amounted to fifty shells per launcher, or ten salvos. The gross weight of the ammunition was 112.5 tons.

In view of the situation, Hitler called in July and September of 1944 for an effective short-term increase in ammunition production for 15cm launchers, from 200,000 to 400,000 rounds per month, and at the end of November 1944, since powder was in such short supply, considered an alternative limitation of range to 1 to 2 kilometers, in order to increase the rate of fire. But any possible saving could not balance the loss in performance! On January 14, 1945 Minister Speer reported that, on account of a shortage of steel, only 37% of the need for launcher ammunition could be met.

Left: The cannoneer pushes the last shell into the barrel until it locks in place. Each barrel was ignited in succession via igniting cable, while the spread struts and locked axles gave the launcher a firm foundation. The shell is three times as heavy as that of the 15cm Nb.Wf. but carries four times the explosive charge. There is no smokelaying ammunition.

A look into the five loaded barrels of the 21cm Nb.Wf.42 with its heavy ignition caps shows the whole threatening might of the medium launcher, which — brought to the right elevation by a simple toothed arc — now awaits ignition. The launcher is basically an aggregate.

A 21cm launcher battery on the eastern front on May 6, 1944, firing (diglycol) — it fired a salvo in eight seconds, two salvos in five minutes.

A 21cm Nb.Wf.42 set up to fire; the ignition cable is already laid out.

After the first large-scale use of launchers when the Army Group Center broke through the Stalin Line in the summer of 1941 and the 11th Army penetrated to Crimea on October 18, 1941, the first unification of seven units (Heavy Launcher Regiment 1, Launcher Regiment 70, Launcher Units 1 and 4) at the focal point of an army with unified tactical leadership by the Commander of Smokelaying Troop 1, Oberst Niemann, took place on May 8-15, 1942 and led to the smashing of the Russian front at Kertsch. This resulted in the formation of motorized launcher brigades, with two launcher regiments, each with three launcher units of three batteries each, for use in German or enemy large-scale attacks; otherwise the firepower would have been excessive. The I. to IV. Units had 72 15cm launchers, the V. Unit had 18 21cm launchers, the VI. Unit had 18 30cm launchers, a total of 108 launchers with 630 barrels. A 15cm unit could also be part of an Armored Weapons Battery (Sfl.). The manpower was 2993 soldiers, the equipment included 109 towing tractors and 284 trucks. The Launcher Brigade (mot.) had the same strength but only half the mobility and 1/6 of the ammunition; the Heavy Launcher Brigade — weaker in terms of caliber and barrels — had the full supply of ammunition but only 15% mobility; for example, St.Wf.Brig.300 on the Oder in 1945. It offered a very effective firepower that was feared by the enemy, especially because of the fast concentration of fire and the shot weights of 35 to 127 kilograms.

At night the shot tracks of the 21cm projectiles were illuminated only during the burning time of the propellants. The ammunition — with only one charge — can be used from -40 to +50 degrees. Because of their thin cartridge walls, their shrapnel effect was meager, but their shock-wave effect was very great, so that a legend of the "compressed air shell" was created by the troops and the Russians threatened in 1941 to use poison gas in reply.

It was necessary, especially after firing very smoky salvos, to change positions with the help of the 3-ton halftrack towing tractor (Special Vehicle 11/1) before the enemy returned fire.

28/32cm Nb.Wf.41

With this heavy launcher, which made the establishment of heavy regiments possible, the principle of bundled, thin, non-rifled launching barrels was given up and launching containers with six rails were introduced that could be adapted to fit various calibers by inserting liners. In this way 28cm explosive shells were used in action, as were incendiary oil projectiles of the normal 32cm caliber. The launchers, very simple in structure, lost any semblance of being artillery ordnance and were improved.

Left: The cannoneer holds the shell, which is 1.19 meters long, weighs 83 kilograms, contains an explosive charge weighing 50 kilograms and is still lacking an igniter, in which state it was taken to be loaded into the launcher, which cost 1835 Reichsmark in all and clearly shows the influence of the 28cm launcher.

28 cm Wurfkörper (Spr.)

28cm Mortar Shell (Explosive)

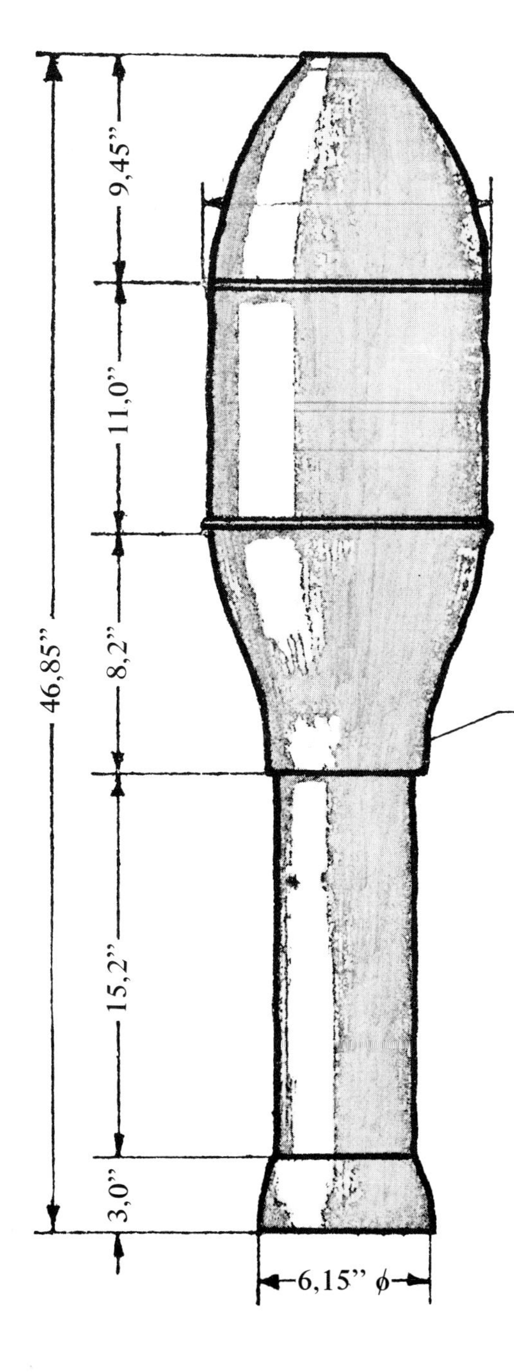

Overall length	119cm
Diameter of charge	28cm
Diameter of propellant	14cm
Weight	83 kg
40/60 explosive charge	36 kg
Propellant: charge	6.6 kg
Barrel length	41cm
Barrel diameter	12cm
Central bore diameter	2.3cm
Ring bore diameter	1.0cm

8 grooves outside the propellant, with celluloid tubes in them

Attachment screw

Bottom view (turbine)

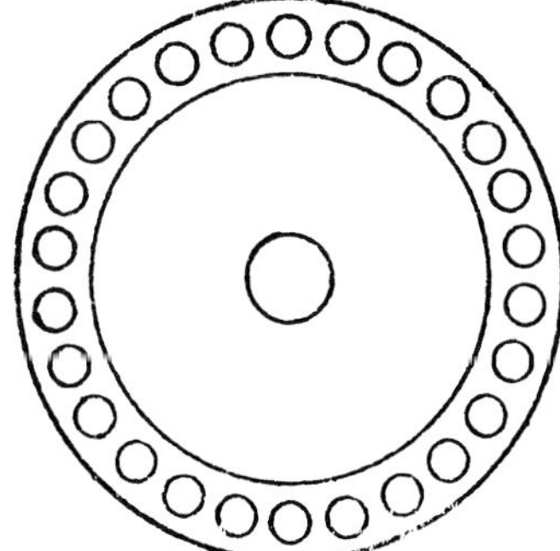

26 jets 0.55" diameter, inclined 12 degrees

The 40 o.V. stick grenades are put in position and the ignition cable for central ignition is unrolled.

Left: Since the ammunition is not impervious to shots, it much be kept away from enemy fire, especially machine-gun fire, what with its short range of 1925 meters.

32cm Shell MFl. 50

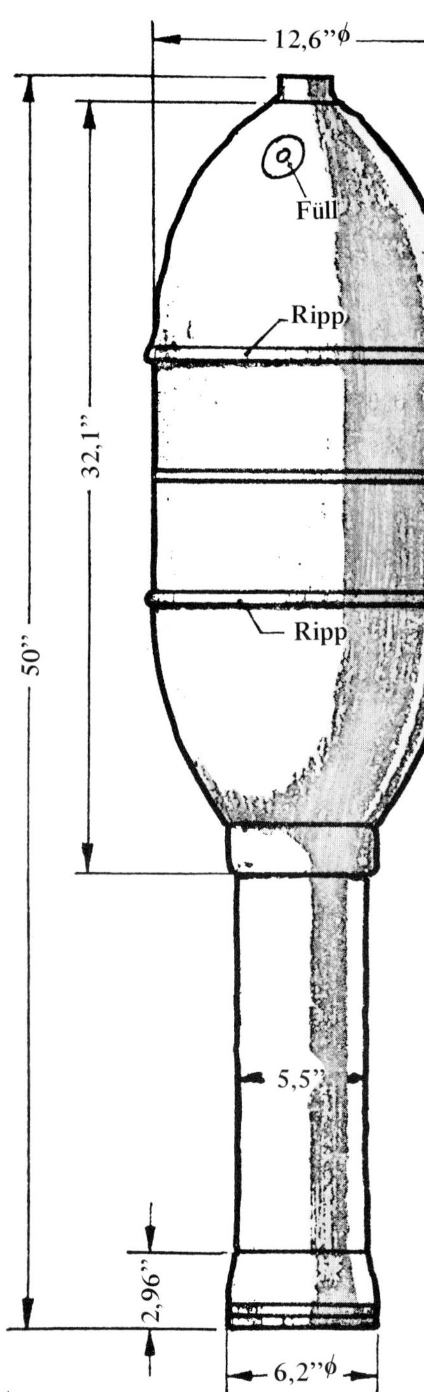

Zdr. Wgr.Zdr. 50 *
+ Gr.Zdlg c/98 Np

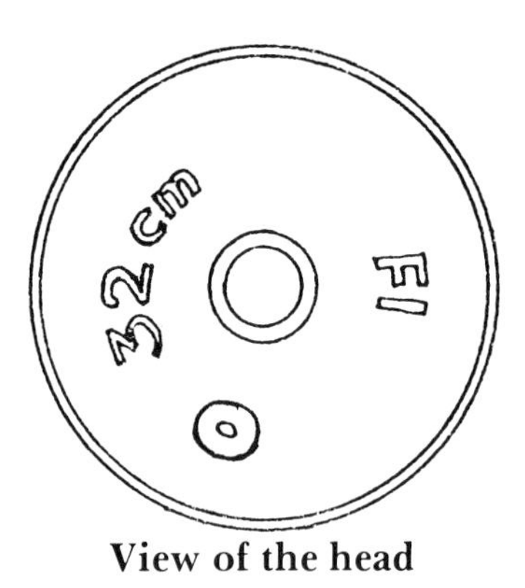

View of the head

Overall length	129cm
Length of charge section	83cm
Length of propellant	46cm
Greatest diameter	32cm
Weight	79.2 kg

The incendiary shell was filled with 45 liters of inflammable oil that had a dangerous effect, particularly when used against habitations. The original equipment of a 28/32 heavy unit included 450 28cm explosive and 150 incendiary shells, a total of 33 rounds per launcher, or five and a half salvos. This equipment weighed 69 tons, of which every battery transported 13.8 tons and the light column 27.6 tons. The explosive shells had a rose red stripe, the incendiary shells a yellow-green stripe on the usual gray-green paint of the shell. As a rule, use of the incendiary shells was combined with that of explosive shells.

The launchers were made by six machine-shop firms in all: Hartmann, Frama, Sack, Eberhardt, Schwartzkopf and Donauwörth Machine Factory in Berlin-Lichtenberg, Hainichen and Chemnitz, Saxony, making four localities in the Reich. In 1944 Skoda became a leader in their manufacture and suggested a compressed air launcher.

The loaded 30cm Nb.Wf.42, when seen from the front, shows the simplicity and extreme economy of the design that was used for explosive and incendiary shells of the same caliber. The previous range was more than doubled without increasing the weight significantly, while at the same field of fire the initial velocity was increased more than one and a half times — a very effective and powerful weapon! With a charge weighing 45 kilograms, more than 1/3 of the 127-kilogram shell is explosive.

Smokeless firing of 30cm shells, "Spreng" (explosive) or "Flamm" (incendiary), from a single-axle Type B trailer. In this further development of the 28/32cm Nb.Wf.41, shot and launching control are identical. The rate of fire is determined by the speed of ignition.

The launcher is now nothing but a portable multiple launching rack, its weight of 1.1 tons scarcely twice as heavy as the 21cm Nb.Wf.42. In the foreground are the packing cases, in the background the gigantic detonation clouds.

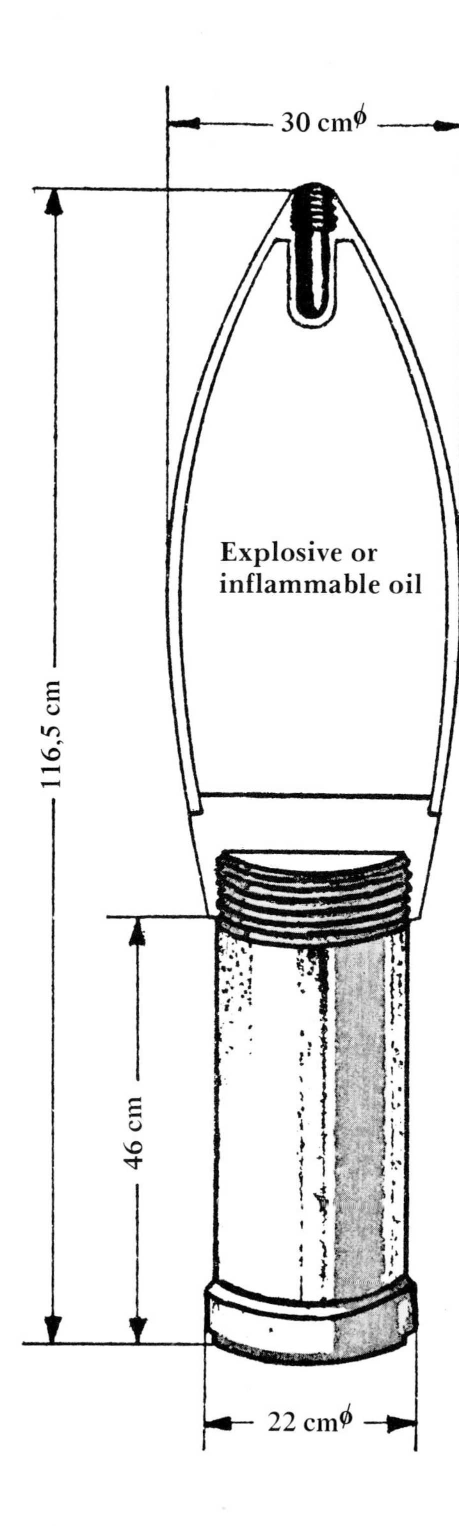

Weight	126 kg
Length	118cm
Greatest diameter	30cm
Propellant diameter	22cm

Parts:
Body filled with 60/40 =66.8 kg
light Jgr. Igniter 23 n/A

Ignition charge 36 Np.
Propellant with turbine, filled = 58.5 kg
Overall length = 57.4cm

18 jets (9.3mm diameter, 12 degree 42 minute inclination)

Weight of Propellant charge 15 kg

(Nitrocellulose 59.9%, diglycolnitrate 35.4%)

Dimensions: (Barrels) 46 x 6.2/0.8cm

30cm Shell (explosive)

The shell design shows maturity, with fewer but considerably enlarged jets at almost the same inclined position as those of the 28/32cm launcher. The electric ignition screw 23 n/A. is used for ignition. The propellant charge is smokeless, consisting of diglycol with nitrocellulose. The effect of the launcher depends less on its construction than on the very effective ammunition!

The heavy launcher unit (mot. 30cm) originally was equipped with 600 explosive shells, making about 33 rounds per launcher in about five and a half salvos. In each battery there are 120 rounds weighing 126 kg, in a light column 240 rounds, with a gross weight including packing cases of some 90,000 kilograms.

In 1944 the 30cm Rocket Launcher 56 was developed and introduced. It fired 30cm 4491 explosive shells weighing 127 kg and also 15cm shells; it cost 3035 Reichsmark in all, and the Donauwörth Machine Factory made fifty of them per month.

Heavy Shell 40 and 41

The mortar-shell launchers were the forerunners of all 28/32cm and 30cm launchers and fired their explosive or incendiary shells from simple wooden or steel frames or even improvised racks, right out of the packing case, which weighed 30 kg when made of wood and 20 kg when made of steel. The Heavy Launcher 40 was made of wood. The fixed-position devices were usually set up in horizontal rows of four in their firing frames. About 40 launching devices with 160 units were installed per battery, which took a lot of time, and ignited electrically by the electric igniting apparatus in thirty seconds.

The heavy launching rack is made of wood. These fixed-position devices were usually used in horizontal rows of four. Some 40 racks per battery, with 160 launchers, were installed at great cost of time and ignited electrically by the electric igniting apparatus in thirty seconds.

In the photo, four heavy Type 40 shells are at right and five heavy Type 41 shells of steel tubing are in position on makeshift launchers. The soldiers nicknamed them "Stukas on Foot" or "Howling Cows."

After the Heavy Launching Device 41 was delivered by truck columns for use, the time-consuming digging in of the containers took place, which determined the traverse aim.

The field of elevation, in which the launching angle was set in the simplest way with Angle Gauge 35 or a pendulum angle gauge, was between 180 and 800 lines. The devices were cheap; a heavy launching rack cost 298.70 Reichsmark, the packing case 41.60 or 45.50 RM, the explosive shell 22.50 RM, the incendiary shell 33.12 RM. The manufacturer was the firm of J. Gast in Berlin-Lichtenberg. A launching rack required 39 work hours, the packing case seven, and the shells themselves 15.8 or 11.1 work hours.

The salvo of a fixed heavy launcher battery sent 8000 kilograms of explosive or inflammable oil at the enemy at one stroke. The range was half that of the smokelaying launchers, with the combat range between 800 and 1200 meters.

Right: Launching Rack 41 is attached to Electric Ignition Chain 40 n.V. before it is ignited. It can also fire 30cm explosive shells, which are fired with C/23 ignition screws and an igniting machine.

Installation of Heavy Launcher 40 (below) and Heavy Launcher 41 (right), loading with 32cm incendiary shells and setting up on a steel pipe frame in 1944.

The 30cm explosive shell had more than twice the range, namely 4550 meters.

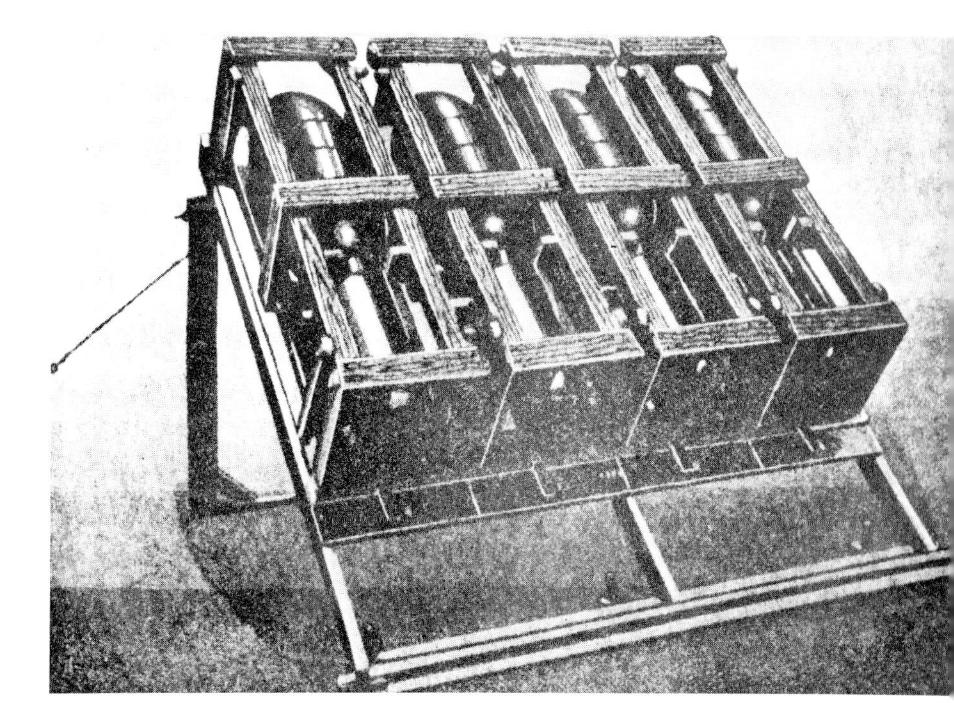

Heavy launchers firing explosive and incendiary shells.

Heavy Launching Rack40

Mounted on the sides of a halftrack and fired from it, this combination allowed quick positioning and thus the necessary change of positions. The soldiers called this halftrack the "Stuka on Foot" or the "Howling Cow." The Panzer engineer companies of the Panzer divisions were equipped with them.

15cm Panzer Launcher 42 (Sf)

Early in January of 1942 Hitler called for a monthly production of twenty 15cm Armored Launcher 42 (Sf.), which was used in front tests as of the autumn of 1943 and proved itself there. It carried a tenfold launching rack on an armor-plated Ford two-ton "Mule" halftrack (Special Vehicle 4/1).

An armored launcher battery halts for technical reasons while on the march.

Unlike the other launcher batteries with six launchers, the armored launcher battery, usually independent, had eight launchers in two columns, which could also be deployed separately. The launcher crew numbered three soldiers.

As of May 14, 1944 the "Pz.Wf.42 (15cm Nb.Wf.41) on self-propelled mantelet" was finally introduced throughout the army. The troops had an urgent need for this weapon, which weighed 7.1 tons and whose 68 HP motor gave it a top speed of 40 kph.

Two-way traffic and horsedrawn units slow the march to the front. With a full load of 80 liters of fuel, the launcher had a range of 130 km on or 80 km off the road.

One at a time, the launchers with their doubled five-barrel launchers move into the darkness from their assembly position to the firing positions they have been ordered to take.

The armor plate measured 8mm all around, 10mm on the turret — similar to the medium SPW (Special Vehicle 251).

The individual launchers have reached their positions — the cab and the launcher at the rear are seen clearly in silhouette. Ardennes, 1944.

The 15cm block of barrels as seen from the front: Turret optical equipment with Aiming Gear 35 can be recognized easily, as can the mounting of the barrel block.

With explosives or smoke, the longest range was 6900 meters and the most effective was 4000 to 6500 meters. The limitation to a load with an initial velocity of 340 meters per second simplified loading and firing so that in 12 seconds the individual launcher could fire ten shots, the battery salvo 80 shots in 18 seconds, and three salvos in five minutes. The main targets of the armored launcher batteries were enemy concentrations, their assignments were destruction by fire and forming of focal points using fire. The battery was excellently suited to be a moving battery, but not a tank or assault weapon!

Loading the launcher with combined power — the elevating machine and electric ignition cable are easy to see.

When loading the launcher, each of which carried twenty shells (two salvos) and had an original lot of 400 rounds in the battery, 10% of them smokelayers, the whole crew is kept busy.

The weakness of the multipurpose vehicle: climbing capability 12 degrees, ground clearance 0.29 meters, fording capability 0.44 meters, 83-link tracks, price of about 22,000 Reichsmark, was in its change of function from a towing tractor to a carrier, in the overloading of the 3.8 ton Carden-Lloyd running gear by armor plate, turret mount and ten shells adding up to 3.3 tons, so that the fighting weight was 1.3 tons higher than normal. This resulted in technical breakdowns, and in addition the armored launcher had only half the range of the medium SPW.

Above: A launcher firing.

Left: Loading the launcher from behind — the launcher leader and cannoneer wear the same field-gray special uniform worn by the assault artillery, with wine-red service-arm color and field-gray beret, their other clothing and equipment being those of the motorized troops.

Opposite page:
Already loaded, the armored launcher rolls to a new position for its next action. The launcher is six meters long, 2.20 meters wide and 2.50 meters high with its complete superstructure. ▶

The battery officer's command to fire unleashes the salvo of an armored launcher battery; the first shell has just left the launcher. Eifel Mountains, December 1944.

8cm Multiple Launcher

A preview before Field Marshal Rommel, Commander of Army Group B, Generalleutnant Speidel (looking downward), and the generals, presented at the Atlantic Wall in the spring of 1944 by a launcher training unit of the army.

Early in March of 1942 the Waffen-SS imitated the Russian "Katyushka" with an 8cm multiple launcher with 24 double rails — firing 48 shells in one salvo. The design was very successful, but it was not introduced in place of the 15cm Nb.Wf.41 because of production factors (machines, work force, ammunition).

Resembling the 15cm Armored Launcher 42, this wide launching rack swivels on a two-ton halftrack, presumably of French origin, even though the launching rack clearly resembles the "Stalin Organ." Dornberger decided against series production in 1942, since the Russian rocket development had remained at the 1934-36 level and stayed some distance behind German rocket development.

Loading the 8cm multiple launcher (48-fold) with shells stabilized by fins. In March 1943, for example, the SS Launcher Battery 521 was equipped with ten 8cm multiple launchers; it later became SS Launcher Unit 506.

Row Launching — Firing Device RG 16

After an already successful performance at the Hillerslebel firing range in 1943, Major Becker's construction command, within the framework of "Action Feuchtinger", presented the "Row Launching-Firing Device RG 16" for additional supply to the units in the west in the spring of 1944; its twenty barrels are seen here at a 60 degree elevation. The device, with an inherent weight of 2.3 tons, could be mounted on a turntable on any vehicle. In this picture, the same French two-ton halftrack off the coast along the "Atlantic Wall" in the spring of 1944.

The multiple launchers' rocket salvos fall just short of the coast along the "Atlantic Wall" in the spring of 1944.

In 1945 a single-load 50cm launcher with a barrel length of 15 meters and a range of 210 kilometers, mounted on a "Tiger" tank chassis, was being developed, as was a two-load 85cm launcher with the same barrel length but, at 80 tons, twice the weight of the 50cm launcher. This clearly pointed the way to modern medium-range rockets. On the other hand, since the spring of 1942 rocket shells, propelled by powder, were introduced for individual guns of the artillery: 15cm Rocket Grenade 1.940, 28cm Rocket Grenade 4.331 for the K 5 (E), a 38cm rocket shell for the assault mortar on a "Tiger" chassis made by Rheinmetall-Borsig as of February 1943, a development that is once again of interest today. Finally the rocket made its way into the antitank weapons — for example, the "Panzerfaust", "Panzerschreck" and the 88mm Rocket Launcher ("Puppchen") — and into anti-aircraft use: In the autumn of 1943 mass rocket action at elevations up to 1000 meters was used against low-flying planes in the Ruhr area; as of August 22, 1943 the use of a special version of the 21cm rocket with an increased shot height from 7000 meters to 14 kilometers was planned. In May and June of 1944 the remote-controlled Schmetterling, Enzian, Rheintochter and Wasserfall anti-aircraft rockets were ready for action. Modern anti-aircraft defense is unthinkable without rockets. Today there is a broad spectrum of rockets in terms of size, drive, technology and use, playing a major role in present-day warfare.

The personnel structure of the German smoke and rocket troops, four of whose five Inspector Generals or Generals of the Smoke Troops came from the Artillery, as set up in mid-1929 and made independent in 1935, remained small in comparison with other service arms, with 5257 officers, 18,150 non-commissioned officers and 88,914 enlisted men, a total of 112,321 men. What they have accomplished is shown by the number of high honors they have gained: nine Knight's Crosses, 64 German Crosses in gold, six honor roll badges and six certificates of recognition, surely unusual for combat support troops of that kind.

The Weapons of the Smoke Troops, 1934-1945

Weapon	10 cm NbWf 35	10 cm NbWf 40	11/18 cm Rauchsp Gerät	15 cm NbWf 41	15 cm PzWf 42
Barrels/Length	1/1,34 m L/13	1/1,858 m L/17,7	1 Rinne, z.4 geb./3,2 m	6/1,3 m L/8,6	10/1,3 m L/8,6
Launchers per battery	6	6	18 i. Erprob.	6	8
Vehicle/Launcher weight kg	111/105	892/800		590/540	7100
Range/Velocity m/sec	3250 m/193	6350 m/310	4500 m	6900 m/340	6900 m/340
Crew	1/4	1/4		1/4	1/2
Shot weight kg	3,7 sp. 7,38	8,9/8,65	Rak(SchwPlv.)	34,15	34,15
Ammunition/Loads	Nb.Spr./3	Nb.Spr./3	Nb.Spr.	Nb.Spr./1	Nb.Spr./1
Shots per minute	10 - 15	8 - 10		Salve 10 sec	Salve 18 sec
Effect	wie Wgr. 37	wie Wgr. 40	Versuche	80 m, Druck(AZ)	80 m, Druck(AZ)
Mobility	3 Trglst/Karre	Laf.m.Bd.Pl.	Zw.Bein m. 60° Schwk.Ber.	Einachsanhgr.	SPW-Maultier 3 t
Notes	ob. Wkl.Gru. Vorderldr. sp. GebNbAbt	ob. Wkl.Gru. Hinterldr.	Rd.Düsen Anzdr. 34 m. Zd.Masch.(sp.Do-Wf)	Sprz.Laf. aufgeb. Digl. Trbldg. vorn	40 km/h. Bttr.Eins.

Weapon	21 cm NbWf 42	28/32 cm NbWf 41	30 cm NbWf 42	30 cm RWf 56
Barrels/Length	5/1,3 m L/6,19	6 Rinnen	6/2,19 m L/7,3	6/2,19 m L/7,3
Launchers per battery	6	6	6	6
Vehicle/Launcher weight kg	605/550	1130/1630	1100/1800	1104/1735(1175 b. 15 cm)
Range/Velocity m/sec	7850 m/320	1925 m Spr/2200 m Fl./145	4550 m/230	4550 m/230(6900 m/340 b.15 cm)
Crew	1/4	1/6	1/6	1/6
Shot weight kg	112,6	82 Spr./79 Fl.	127 Spr./Fl.	127 (34,15 b. 15 cm)
Ammunition/Loads	Spr./1	Spr. 50/Fl. 41,5 kg/1	Spr./Fl. 50 kg/1	Spr. 50 kg/Ng. wie 15 cm/1
Shots per minute	Salve 8 sec	Salve 10 sec	Salve 10 sec	Salve 10 sec
Effect	100 m, Min. u. Hochdruck	ger.Spli.hoh.Druck	Druckwelle	Druckwelle
Mobility	Einachsanhgr.	Einachsanhgr.	Einachsanhgr.	Einachsanhgr.
Notes	AZ/mV, zugl. 15 cm Wf.Gr., Sprz.Laf., Trbldg. verteilt	ab 15. 7. 43 aufgegeben	Ende 44 aufgegeben	ab Ende 44 EinheitsWf. f. 15 cm u. 30 cm

Weapon		sWuG 40	sWuG 41	sWF 40
Barrels/Length		4	4	6
Launchers per battery		40	40	je mSPW(SdKfz 251)
Weight of frame/case kg		52/30	110/20	/30 od. 20
Frame/firing position kg		488 Fl.	548 Fl.	7654 Fl.
Shot range	28cm	500 Spr.	558/738 Spr.	7672/7942 Spr.
	30cm	1925 m Spr.	dto.	dto.
	32cm	- Spr.	4450 m	dto.
Velocity m/sec	28cm	2200 m Fl.	dto.	dto.
	30cm	145	145	145
	32cm	-	230	230
Shot weight kg		145	145	145
Ammunition/Loads		82 Spr./79 Fl.	82/127 Spr./79 Fl.	
Shots per minute		Spr.Fl./1	Spr.Fl./1 Spr.Fl./1	
Effect		Salve 6 sec	6 sec	10 sec
Mobility		hoh.Druck	Druckwelle	
Notes		Stell.Wf. WG Holz Vortrb. d. 26 kl. Düsen	Stell.Wf.Sf WG Stahl	Pz.Pi.Kp.

Wgr: Wurfgranate = smokelaying shell;
NbWf: Nebelwerfer = smokelaying launcher;
PzWf: Panzerwerfer = armored launcher;
RWf: Raketenwerfer = rocket launcher;
WF: Wurfrahmen = launching frame.

Development of the German Rocket Launcher

1927	Development of Army Gunnery School in Celle.
1929	Mid: Army Weapons Office (HWA) tries to set up smoke columns in infantry & cavalry divisions indicate need forspecial troop and lead to restructuring of 2./Fahr-Abt.4ting smokedevices & developing jet weapons (black-powder weapons).
1930	Systematic testing of all smoke devices, as of summer mil. testing of large-caliber rockets by HWA under far-seeing leadership of chief, Gen. of Art. Prof. Dr. Ing. Becker, end of year, beginning of systematic testing of powder rockets by HWA at Kummersdorf firing range, first rocket free-flight experiments.
1931	Development of spin jets by HWA.
1931/32	Tests with smoke pulverizers and powder rockets, plans for long-range rockets.
1933	Summer: transfer of 2./FA 4 as 4./Fahr-Abt.4 to Königsbrück, Prinz Georg Artillery Barracks.
1934	10/1 renamed "Art.Abt. Königsbrück" under Maj. Ochsner w/ 2 batteries: 1./Hptm. Maltzahn, 2./Hptm. Dr. Dornberger, with 10cm NbWf 35, 2./ also w/ 11cm "Tracer Device", 3.20 m. long, as troop test (first rocket launcher).
1935	Readiness of new weapon first proved at Jüterbog firing range, 10/15 unit divided, renamed "Nebel Abt.1 Königsbrück" (Staff & 2./ under Maj. Knecht) & "Nebel Abt.2 Bremen" (Staff & 1./ plus Lds.Pol. & Cav. under Maj. Ochsner, 18cm "Tracer Device" for troop test w/ 2./2.
1936	HWA/WA Test 11 for all army rocket developments April: setup "Inspection of Smoke & Decontam. Troops (OKH/In 9)", the Smoke Troops become a service arm (wine-red color), 10/6 "Smoke Training & Testing Unit Bremen" set up (of Smoke Units 1 & 2) under Maj. Kanzler w/ Test Staff Maj. Maltzahn w/ tasks: 1. firing tests, 2. decon.
1937	Improved 15cm "Tracer Device" w/ Smoke Training & Testing Unit (Do-Device), 10/1 transfer of unit under Oberstlt. Knecht to Celle, 10.12 3rd batteries set up in all smoke units, 15cm NbWf 41 developed as superior standard weapon (multiple rocket launcher).
1938	10/1 "Smoke Unit 5 Münsingen" set up under Oberstlt. Grothe, testing of 10cm NbWf 50, unsatisfactory results.
1939	Smoke rocket production restructured as highest production of explosive rockets, 9/1: 4 smoke units exist, Nbl.Abt.1 serves w/ Army Group South, Nbl.Abt.2 w/ North, "Nbl.Abt.3", Nbl.Ers.Abt.1 & Entgift.Abt.101-103 set up in Celle district, followed later by 18th Ers.Abt. & 1st Ge.Nbl.Wfr. Ausb. & Ers.Abt, October: split of Nbl.Wfr. & Entgift-Abt.
1940	1/1 "Nbl.Abt.4" set up, black powder propulsion dropped,February: Smoke Training School set up in Celle (of former test staff), 4/18 first demonstration firing by 1st Smoke Training & Testing Unit w/ 8 15cm NbWf 41 in Munster before Army Cmdr. Generaloberst v. Brauchitsch & Genlt. Ochsner, General of Smoke Troops announces rocket weapon is ready for front use, Nbl.Ers.Abt.2 Bremen set up, all 6 Smoke Units with 10 NbWf 35 in French campaign. 7/10 first Launcher Reg. 51 set up (of 2./Nbl.L. & V. Abt. & cadres) Munster w/ 15cm NbWf 41 (324 barrels), in autumn Launcher Reg. 52 Fulda, Launcher Reg. 53 Bocholt/Bottrop, Launcher Reg. 54 Posen, September: Smoke Training Reg. set up (of Nbl.L. & V.Abt.) under Oberstlt. Maltzahn, auxiliary use of Heavy Launcher 28/32cm in Decontamination Unit.
1941	4/18 "Werfertruppe" independent, 7 smoke units rearmed with 15cm NbWf 41, 6/22 4 Launcher Regs. & 1 unit Smoke Trng. Reg. in surprise front service in Russia, Heavy Launcher Reg. 1 set up in Celle/Uelzen w/ 28/32cm NbWf 41 (devel. as Heavy Launch Frame), later 30cm NbWf, 28/32cm NbWr 41 in front use, other new launcher regs. set up, 28/32cm Heavy Launch Frame in service.
1942	Very successful large-scale use at Kertsch ("Trappenjagd")and in capture of Sebastopol ("Störfang"), Smoke Training Reg. 2 set up, 21cm NbWr 42 introduced.
1943	Spring: Regiments restructured—launcher reg. now consists of 2 light, 1 heavy unit, heavy launcher reg. of 2 heavy (28/32 or 30cm) units, 1 light (15 or 21cm), Summer: Launcher brigades formed (Cmdr. of Smoke Troops), 30cm NbWf & 15cm PzWf at front, launcher units in Waffen-SS.
1944	October: Smoke Trng. Regs. renamed "Launcher Trng. Regs.", December: greatest launcher mass von Düren to Pfalz for Ardennes offensive as of 12/16.
1945	Launcher troops include 17 brigade staffs, 50 launcher reg. staffs, 16 of them heavy launcher regs, 150 launcher units, 46 smoke units, 13 Pz launcher batteries w/ 4816 launchers, 5 decontam. units, 18 Nbl.Ers.Abt.w/ 27,066 vehicles, 3/27 Smoke Trng. Reg 2 Munsterlager w/ 7491 men transferred to the Weser (Div. No. 480, Corps Ems/AOK Blumentritt). Army Noncom School for Smoke Trng. Celle within Div. No. 471 transferred to west end of March. 4/11 complete destruction of Donauwörth Machine Factory (production facility for heavy launchers) by air raid. 5/6 Taking of Pilsen by V. US Corps puts Skoda Works, leading in German launcher production, out of action.

„UHU"-He 219
THE BEST NIGHT FIGHTER OF WORLD WAR II
SCHIFFER MILITARY

ARMORED MILITARY VEHICLES
MAUS
AND OTHER GERMAN ARMORED PROJECTS

DORNIER DO 335
"PFEIL"
THE LAST AND BEST PISTON-ENGINE FIGHTER OF THE LUFTWAFFE

GERMAN
ARMORED TRAINS
IN WORLD WAR II

SCHIFFER MILITARY

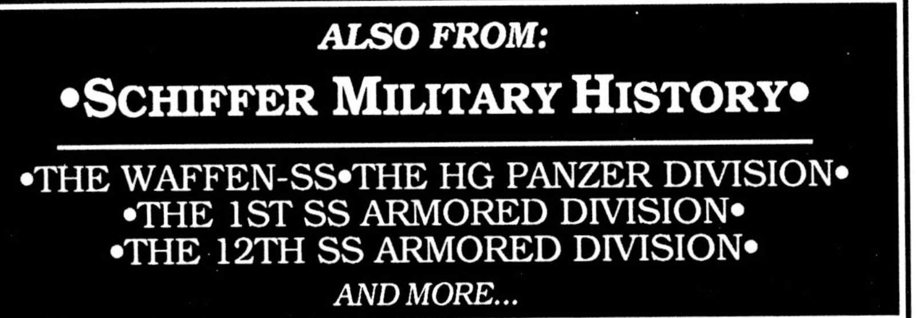
ALSO FROM:
•SCHIFFER MILITARY HISTORY•
•THE WAFFEN-SS•THE HG PANZER DIVISION•
•THE 1ST SS ARMORED DIVISION•
•THE 12TH SS ARMORED DIVISION•
AND MORE...

THE WORLD'S FIRST JET BOMBERS
ARADO AR 234
JUNKERS JU 287
by FRANZ KOBER
ARADO AR 234 B-2

GERMAN AIRSHIPS
PARSEVAL-SCHUTTE-LANZ-ZEPPELIN
HEINZ J. NOWARRA
-SCHIFFER MILITARY-

GERMAN BATTLETANKS
IN COLOR 1934-45

GERMAN NIGHT FIGHTERS IN WORLD WAR II
MANFRED GRIEHL
-SCHIFFER MILITARY-

The King Tiger Tank
by Horst Scheibert

THE PANTHER FAMILY
by
HORST SCHEIBERT

GERMAN PERSONNEL CARS

CAPTURED TANKS UNDER THE GERMAN FLAG
RUSSIAN BATTLE TANKS
-SCHIFFER MILITARY-

GERMAN SHORT-RANGE RECONNAISSANCE PLANES
1930-1945